YOUR KNOWLEDGE HAS VALUE

- We will publish your bachelor's and
  master's thesis, essays and papers

- Your own eBook and book -
  sold worldwide in all relevant shops

- Earn money with each sale

Upload your text at www.GRIN.com
and publish for free

Adaptive Features and Context Awareness of a Smart Car

Hardik Modi

Karna Joshi

Palak Patel

Bibliographic information published by the German National Library:

The German National Library lists this publication in the National Bibliography; detailed bibliographic data are available on the Internet at http://dnb.dnb.de.

ISBN: 9783346762306
This book is also available as an ebook.

© GRIN Publishing GmbH
Nymphenburger Straße 86
80636 München

Print and binding: Books on Demand GmbH, Norderstedt, Germany
Printed on acid-free paper from responsible sources.

The present work has been carefully prepared. Nevertheless, authors and publishers do not incur liability for the correctness of information, notes, links and advice as well as any printing errors.

GRIN web shop: https://www.grin.com/document/1268378

AN OVERVIEW ON ADAPTIVE FEATURES AND CONTEXT AWARENESS OF SMART CAR

Karna Joshi, Hardik Modi, Palak Patel

V.T.Patel Department of Electronics & Communication Engineering, Chandubhai S Patel Institute of Technology, Charotar University of Science and Technology,Changa,Anand,Gujarat,India

Abstract

Smart car is future of developing world. With advancement in research in different sectors of car features that can provide a safety and security to costumers. This paper comprises of automatic cleaning system robot, accident detection technology in car, smart car parking features. In addition to this automatic cleaning system robot, smart car parking and accident detection technology all comprises of involvement of IoT (Internet of Things). In keen car programmed cleaning robot it can be worked naturally as well as physically. In mishap location as before long as the car identifies the mischance flag from sensors are consequently received by cloud and alarm message is sent to individual who is subscribed to car. In smart car parking system the customer will be assisted how to park the car and help to find the parking space . In expansion, it has the traffic status features which will help the people to reach destination easily . In addition, it collects the data from the cloud and then by the help of data filtering it will help reducing data over network.

Keywords: IoT, GPS,Automatic Cleaning System Robot(ACSR),Smart Car Parking , Accident Detection Technology

Table of Contents

Introduction

As we are getting used by new technologies in recent years there is many technologies that has changed the future of car. There are many IoT projects going that can change the future of car and we can refer to as future smart car there can be many technologies combined and can be future of cars. Basically, according to some research an individual spends average 45 minutes in a car[1]. There are many ways that can make a smart car future of industry . In today's society there's a IoT could be a term utilized to communicate between individuals . Businesses are contributing exceptionally extra due to various advantages of IoT in different areas. In current situation there is rapid increase in car industry it used to be 841 million cars in 2008 but till 2035 there would be 1.6 billion cars approximately.

There are many new innovations going on in field of IoT where the engineers are working on making the future smart car there is a rapid change in increasing number of cars so nowadays there are many circumstances where the car doesn't get the desired space for parking in the public places such as stadium, hospitals, shopping malls etc. Finding unoccupied parking space for a car is vast problem nowadays [2]. There is multiple parking lanes and the biggest issue of all is air pollution when the car doesn't get enough space for parking and needs to wait. The researchers have studied on use of IoT in the transportation and computing[4]. A few models have been implemented and that gives the information on the available parking slots [5]. Also, a few of them proposed arrangement to gather and send the information to the cloud preparing center which decides the arrangements and gives these back to the car parks. In smart car stopping information is been collected by the diverse sensors in indoor and on road stopping Furthermore the information collected and it's been examined and handled by the IoT gadgets. The information will be assessed by utilizing calculations, which in turn forms concurring to the provided conditions. In expansion, the provides portable user interface that helps the consumer effortlessly check the closest car stopping with the distance from user activity by means of Google API, which helps the user with accurate data .The cloud will help and collect the data from fog microcontroller that tracks user location and analysis as well as process the data. Car stopping extend is information combination and sifting the information collected by the sensors. The steps for handling information is collection, channel, meld, preprocessing, putting away and conveying the data.[3] The information sifting work channels the sensor information to decrease the sum of transmitted information. Information fusion integrates sensor information to supply exact information to clients. Information handling is done at the source and information is put away within the cloud. Usually fundamentally an awesome innovation utilizing IoT which spares the time of shopper and makes a difference to discover a culminate spot for stopping.[1-30]

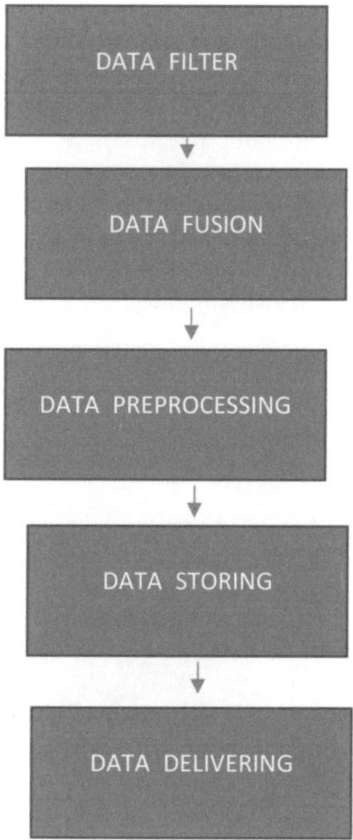

Figure 1. STEPS OF PROCESSING DATA IN SMART CAR

Further about the automatic cleaning robot and the main thing here is the power which is the main constrain while talking about the vacuum cleaner of 200W hence here we need to add the external power source that can be used and also charge the source accordingly. The outlined robot can be worked utilizing IoT and smartphone application which can be direct communication between client and robot. Space is the major issue whereas making the robot. Here is the mental thought of shrewd cleaning robot.

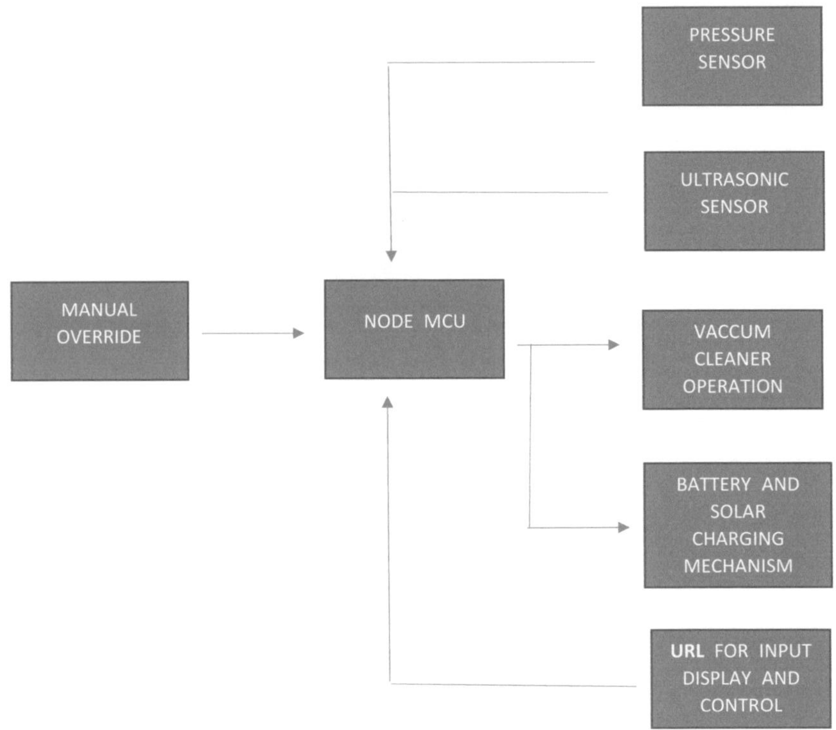

Figure 2. INTEGRATION FOR AUTOMATIC CLEANING

Ultrasonic sensor is the sensor which can be used to calculate the distance between user and the robot . Another main sensor here used is the fire sensor which is used to detect fire if the robot is used under the tough environment. Motion sensor as the name suggest it is very useful for detecting the motion so it can useful for placing the other robot at particular distance this also helps to carry the robots which need to be repaired and maintained .There are many cleaning bots available in market that are ROOMBA(2002) which nearly cost around Rs 35000, SCOOBA (2005) which costs Rs 35000, DC-06(2001) which costs Rs 70000, STRIC (Smart Car Interior Cleaning Robot) which is around Rs 7000. Here we are going to discuss about STRIC.STRIC begins with the kilometer run if it is satisfied then and only then it will take input from the pressure sensor . If the user wants manual friendly system then there is push button provided in the cell phone connected over the internet so it can be switched to the user manual mode .

Here there are sensors two sensors used one is ultrasonic and other is pressure sensor.

Further lets discuss something about accident detection system. There are several technologies that are used that can be vehicle to vehicle communication vehicle to infrastructure communication and other is combination of both the systems .In this IoT system there is a automatic response which is ery useful to the user that does the several thing that is firstly the accident is detected by the microprocessors and the sensors used and then the information is collected from the cloud and then sent to nearby hospitals and the ambulances.Here there is unit of raspberry Pi and that will take the advantage of the position and the location and the accelerometer detects the collision . In this mobile will constantly communicate with the database. By detecting the accident the mobile gets notified and by the help of Google maps API the latitude and longitude of the location is send to the mobile application so the notification is sent to owner's family as well as the authorities.In order to do same the longitude and latitude are sent to the internet and then it will compare the provided details and then it will return the string variable. Typically at that point spared and utilized in caution messages. Longitude and scope upgrades each 5 seconds for the exactness. In case the modern arranges are diverse from the earlier ones at that point the application goes within the running mode and checks the unused area[5-7].

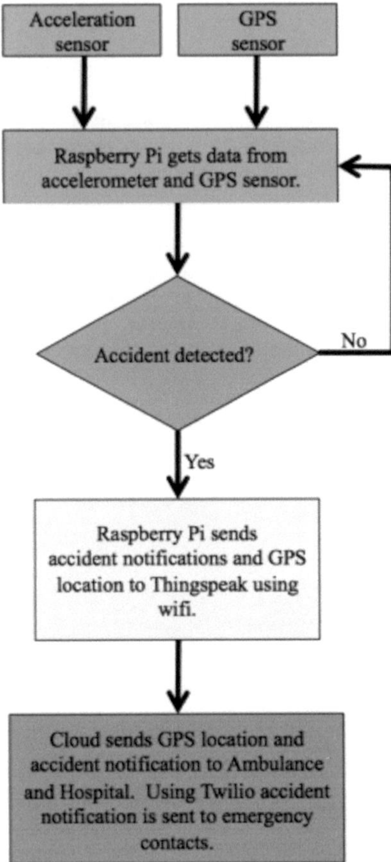

Figure 3. BLOCK DIAGRAM OF ACCIDENT DETECTION SYSTEM

There are many circumstances and problems faced while making all this things possible there are the ways by which this problems can be sorted out. While adapting the advance parking system the fusion and filtering data is generated every second to it is a big question that where and how to conduct the data , processing the data is the important aspect as unused data will not be transmitted so that unused data will be stored in the cloud which can be very expensive and costly. Another challenge faced while making parking system was identification of the machine learning techniques algorithms . The communication between the user and cloud was using HTTP(Hypertext Transfer Protocol) protocol[9].

There were many issues in cleaning robot there must the suitable mechnical condition from car side that must be fulfiled while making the cleaning robot the matter of fact was that the voltage condition is must while working on the robot that is 200W and that can be solved by adding the external power supply. Another constreint is that the weight and size of vaccum cleaner.Ultrasonic sensor is used in the bot. There are many issues that can be faced and there are many solutions to it but at particular time that can be more expensive and can't be afforable.

Futhermore, talking about the updation of the system that is the important thing to be noticed as there are many new technologies introducing day to day basis so accordinly we need to update the technologies and make the product that can be used and implemented in the future. In this paper we have discussed about the different IoT technologies and combining it to create smart product[10].

In beginning, we have discussed about the advanced car parking that is very important aspect of the generation.Nowadays, people prefer to use own vehicle than public transportation. In 2003, it was 907K cars and 253K commercial vehicles. According to reports in Dec 2018 there are 74,502K registered vehicle. In today's generation there are many many spots where there is the use of parking guidence that is very useful for the generation that are used for the unoccupied parking space and then guided.In few studies there are the points where the data of parking slots are collected and sent to user. It can be done by tracking the traffic and using google API the parking location can be found. Smart car parking can be useful and it can be said that more effective process of the century . Some researchers also have studied and guided that all the future car can have built in function of the parking allocation that can reduce the traffic and collisions that are created by not getting proper parking allocations.

Further, discussing about the automatic cleaning robot in current situation generation is very busy and not able to maintain proper hygiene so there is requirement of the smart car interior robot that ca help to maintain the car and also helps in proper hygiene. According to the researchers there is 1.9 million deaths due to pollution in 2019 and in that quarter time by lack cleanliness in the area.

Additionally, talking about the accident detection system the study where the system speed is controlled in the school or the construction areas when the machine is in operating mode the vehicle speed can be operated automatically using RF transmitter. Another great example is the IoT monitoring system for the heart patient for the great risk of myocardical infraction and unusual change in temperature of the body . There are still places where the individuals are curious about the circumstance more than making a difference harmed individuals the IoT based framework will directly contact the family part and than the emergency vehicle so that it'll offer assistance the quiet to reach the clinic as before long as conceivable and than makes a difference in recuperation speedier. The passing rate of individuals due to mischance is exceptionally tall as the vehicle proportion increments it increments the chances of mishap. In 2020, there were about 132K passings due to street mischance[11-15].

A smart car be very useful and can also can be helpful in self driving system. In IoT there are many chances that it lead to the system fail but it can be useful at the same time. At some time of instance lets talk about the IoT processes in self driving system it can be useful and at the end we just say that it is just a metal box with 4 wheels but by just using the IoT functionality it can be a smart car and there are various components in the IoT that can change the future and the technologies that are currently acquired that can be in built advanced car parking facilities and accident detection including can change the future of the car. [15-17].

There are many countries that are just focusing on the smart car technologies among which the highest rated nowadays id United states. Other countries including the UK, Europe and Netherlands are progressing the same. The breakthrough will be to the manufacturers that are manufacturing the cars and vehicles there are autonomous vehicle that are of level 3 and 4 will be manufactured by around 2021 but the person needed the level 5 they need to wait a lot. In upcoming three to four years the major manufacturers i.e. TESLA and more would have autonomous vehicle in sale[8] .

5G the innovation of car

As the car business keeps on speeding up the improvement of associated vehicles, arrangements are arising progresses in systems administration, 5G, and IoT. Customary vehicle producers need to take on extra jobs past simply making vehicles and becoming specialist co-ops for vehicles.

A few details of the 5G standard are as yet being characterized, yet it is now referred to that media communications won't develop as ever before [23]. The technology , with a hypothetical data transmission, diminished utilization, tremendously diminished idleness time, and exceptional unwavering quality, that shall be universe of tomorrow [24]. It can be fundamental for brilliant vehicles. It is fit for utilizing the net at high velocities and a progression of beforehand inconceivable administrations.

Brilliant vehicles, business 5G cell phones, and 5G organizations will ensure better solace, security, and proficiency . This stylish coordinated effort united the force of eye-observing glasses, smart tires, expanded reality locally available the motors, and 5G cell phones in cooperative energy on the 5G organization. All this things can be done by coordinated efforts between vehicles, the IoT environment, and important potential high level help administrations.

By decade in 2030, then, the unrest will have much brighter future. As per scientists of the Hyundai , in the following decade, the vehicle business will be redesigned by lots of megatrends that are now beginning to arise: from associated culture , with the development of in internet of things likewise to vehicles, to the maturing populace, which will expect concentration on vehicles adjusted. Another concern to nature will be megatrend, by development on vehicles followed by accommodating mixture motors , hence, the designated "multi-facet organization"; that is, the blend of various modern areas that will team up with the car area to increase, on account of new advancements in the driving . There is a plausible that give customized data for clients in light of natural and profound signs. Drivers shall have encounter of versatility as to customized in-vehicle administrations adjusted close to home continuously. To draw in and dazzle clients, the auto business will likewise must be progressively imaginative and innovative in proposing vehicle models that are combinations of craftsmanship and innovation, out of reach for normalized inventive calculations of man-made reasoning [25].

The constant mechanical headway could augment uneasiness and turmoil in the public eye, and the vehicle will now not had the chance to be in a manner of speaking be a vehicle of moving between different places, yet besides a put of escape. The add will, consequently, be easy to understand. By and by, simultaneously, the improvement of economy that is shared prompt execution for request portability organizations that stages are augmenting the client reviews. By the development of automated driving vehicles, it can be logically basic to set up moral principles for flexibility structures to guarantee security, control, and viability. Regardless, movement issues will grow dramatically because of the extension of urban areas, with 70% of the world people living in metropolitan zones. The vehicle and flexibility will also must be reevaluated[18].

After a decade, in 2040, vehicles will go through another change. By that date, it is normal that the ignition motors will vanish for the selective utilization of power. Once more as per Mike Ramsey [46], head of Gartner's exploration bunch having some expertise in shrewd versatility and the car business, during a meeting with CNBC, confidential vehicles will turn into a rich superficial point of interest. By the development of economy and the expansion of administration bills, suburbanites will generally utilize the part of portability administrations, or the dire outcome imaginable, to purchase recycled vehicles, and the vehicle will become, in the greater part of the cases, a help (CaaS) and presently not a resource. The further evolved vehicles will rather have go seats to make working spaces inside them, further created HUD screens than those available today, biometric structures instead of keys and locks, orders instituted by signs and voice control, and an intuitive media place that will transform into the focal point of the vehicle's infotainment system (and which will similarly be related with sensors and devices of the adroit home), by which driving will transform into the commitment of the on-board PC ultimately[19-22].

As of late, the consideration of vehicle security engineers has stretched out from inactive wellbeing frameworks, (for example, airbags, safety belts, and effect opposition, to give some examples), which are now merged, to cutting edge dynamic wellbeing rehearses, chiefly intended to forestall risky circumstances and mishaps. High level driver help frameworks (ADAS) are electronic driver help frameworks created to further develop vehicle security [23-24]. ADAS is a condensing that shows and gatherings a progression of gadgets that can help and work with driving the vehicle even in crises. ADAS incorporate different gadgets, for example, the downpour the actuators to actuate the wipers consequently, sundown actuator for the programmed controlling the led lights, versatile journey capability to change the rate in view of traffic, programmed crisis slowing down, stopping sensors, path change cautioning frameworks, and programmed signal acknowledgment frameworks. A significant number of parts have been introduced recently on vehicles and can portrayed more in the accompanying sections.

Park Assist

The recreation area help is outfitted with sensors provided to recognize reasonable for leaving vehicle, equal and opposite to the heading movement, brush [5]. By and large, framework can be enacted with squeezing for a reasonable sign on permitting the gadget begin filtering encompassing area looking for a station spot. The driver can confer to the control unit from which side it is jumped at the chance to come by implanting the reasonable sign. Whenever it is found a spot reasonable for the size of the vehicle, more extensive or longer, the control unit shows a message in plain view. The beginning place of the move is displayed to the owner that draws in the converse stuff once the appropriate spot is adapted. At this point, the system takes command over the guiding and shows to proprietor, through the locally accessible structures, the feature, and voice synthesizer, how to manage the gas pedal or the brake for pivot to stop precisely.[22-23]

Cruise Control

ACC is a convenient gadget, particularly by the roadway, that assists with keeping up with consistent speed however regarding a base preset distance contrasted with the vehicle in front . It is versatile on the grounds that it naturally adjusts the vehicle's rate of the other vehicle going forward. The vehicle along with voyages, for example, at 85 speed rate, and the ACC has adjusted at 145 speed rate an arrangement of actuators, perhaps a lens removes the motor it till the vehicle dials back to a speed of 85 speed rate although a security area from the forward one that stays constant. At point if the vehicle ahead speeds up, from 85 speed rate to 195 speed rate, the vehicle will speed up in the up to the set speed of 145 speed rate. At point when the machine is outfitted by a programmed transmitter, the framework could slow down it until it stops totally.

Traffic Recognition

It is the acknowledgment that can identify example speed restrictions, yet in addition disallowances on access and surpassing. A camcorder, situated within the windshield, registers as far as possible showed out and about edges, yet in addition those on the beam entryways or close to building destinations. The information caught by the lens contrasted and the data of the route framework that showed both on the instruction board show and the instruction screen [26]. Allowed or surpassing boycotts (counting showing end) and different alerts additionally reviewed. The lens, utilized for it for acknowledgment, permits makers that coordinate an extensive variety of driver help frameworks, further developing security ready.

The Future of Diesel

Vehicle producers, mostly determined by the developing worry for environmental change, and part of the way (more everything being equal) by progressively harsh outflows stakes, presently seem, by all accounts, to be checking out solely an motors for vehicles. Diesel motors and none of the petroleum ones, could help in that frame mind of pushing onwards to electrical ones. On the top of that thought, it is feasible to note Bosch, as indicated by which diesel motors dirty even not as much as petroleum motors [26]. Meanwhile, in Europe, a few states have proactively demonstrated vehicles that ignition motors can presently not be sold after 2039, while different nations are thinking about halting and others have even expected the boycott: in Norway 2023 and Holland in 2029 . In any case, Mercedes-Benz specialists have brought up that diesel is the most productive ignition motor, so it will stay dynamic for quite a while — past 2023.The greatest companies , consequently, feature that fuel isn't expected can vanish. Also, the British companies , notwithstanding obligation that power, proceeding that contributes many tons of the advancement of more productive diesel motors with less outflows[25].

In any case, the future brilliant vehicle may not just bring benefits. For example, more incessant disappointments and irregularities should be tended to. For the most part, the more intricate a framework turns into, the more it is leaned to breakdowns and startling issues. Unquestionably, the mediation of exceptionally concentrated specialized work force will be expected to tackle these issues. Moreover, as currently referenced, taking into account that the latest thing is centered around a gigantic shift of exploration and ventures towards the electric vehicle, when there will be a great many electric vehicles, from one perspective, the contamination in urban communities will be diminished. Notwithstanding, it will be important to manage the expansion in energy interest. Thus, huge interests in the inexhaustible shall be required, and there can be an expansion on the interest for capital, principal needed for the development of batteries. Toward finish of the reside, the last option shall discarded try not to turn into a perilous wellspring of contamination[27].

Child Detection system

Youngsters left inside an vehicle that has no owner shall be incredibly perilous. As per the report in [5], a normal of 38 youngsters pass on from brain stroke inside the extremely heated vehicles every time, and the is a sum of 800 brain stroke and heatstroke passings that have been recorded along the British states for the long duration from 1995 to 2015. Among those, 60% of the PVH passings are because of that guardians left their youngster in the vehicle and 40% are because of the kid having accessed the vehicles without telling the guardians. In any case, on the grounds that the greater part of the children are placed inside a vehicle on the secondary lounge and the mass is less, momentum pressure actuator and lens neglect to identify that they are present. Subsequently, it is basic and earnest to foster vehicle checking frameworks distinguish expeditiously at that time youngster is completely locked in that vehicle with the impediment so hence keep away from such a misfortune[28].

Supply of Power In Future cars

Relinquishment in petroleum derivatives that can progress to power high the risk of issues so it couldn't be difficult that address. Dispersion signals should have the option to endure such an enormous energy interest on the off chance that electric vehicles supplant the whole armada of cars.Besides, in the region and unassuming communities, all things considered, homes have a carport for the vehicle (and thusly admittance to a power plug), while in huge urban areas re-energizing the vehicle batteries might be more troublesome. A huge public mediation will be fundamental for improve both energy creation (clearly from sustainable sources; in any case, the advantage is that "green" motor is that it is lost by simply arranging the emanations from the spot purpose to that spot of creation) and the re-energize limit. This framework gives the likelihood to the electric administrator who deals with the segments that operates the vehicles are batteries that are being re-energized so it can be utilized as "brief collectors" to preseve abundance creation and "helper batteries" also ingest into energy be taken care of into the matrix whenever it is necessary[28-29].

Conclusion

In this paper, we have examined different IoT advancements by which a common vehicle can be switched over completely to smart vehicle. We have examined about brilliant vehicle leaving utilizing IoT, smart cleaning robot, mishap discovery framework. The finding from brilliant vehicle leaving framework was decrease of sent information through organization and saving the energy . The finding from cleaning robot was the most client target is vehicle maker it is fundamental that the data that to be given to client should be smaller and all that ought to be kept prohibitive the full construction is inborn such a way that by and large the immense data is passed on the client. the center thought of the expand is free cleaning in any case there's a manual mode which has been given for the expect of the client. It has been made useful with the assistance of a url which can be carried with either a mobile phone or PC or any proposes with which has web connection. Thus, it is important to assess whether sending something similar "directions" to man-made reasoning that might need to conclude in a genuine situation is correct." Regardless, this id not yet the question replied, however last option shall be major, through from legitimate liability and starting along the protection state, when oneself operating vehicles shall be available for huge scope at the roads. Obviously self-driving fender benders are a certainty, yet the range of mishaps that morally harmed people in general and upset the business are not. Subsequently, it seems critical to foresee how society will answer the moral decisions customized into these smart vehicles. Taking into account that any course of action utilized in gathering these choices will have its tendencies and shortcomings, the systemic heterogeneity and the boundless contribution of clinicians will be critical to moving toward that objective.

References

[1]   Mamun, Md Ibrahim, Afroza Rahman, Md Abdul Khaleque, Muhammad F. Mridha, and Md Abdul Hamid. "Healthcare monitoring system inside self-driving smart car in 5g cellular network." In 2019 IEEE 17th International Conference on Industrial Informatics (INDIN), vol. 1, pp. 1515-1520. IEEE, 2019.

[2]   Zhao, Ruoyu, and Minyong Tong. "Research on 3D Mapping Technology of Smart Car based on Depth Camera." International Core Journal of Engineering 8, no. 1 (2022): 467-474.

[3]   Yang, Byungyun. "Developing a mobile mapping system for 3D GIS and smart city planning." Sustainability 11, no. 13 (2019):

[4]   Krishnan, Prabu. "Design of collision detection system for smart car using Li-Fi and ultrasonic sensor." IEEE Transactions on Vehicular Technology 67, no. 12 (2018): 11420-11426.

[5]   Shaik, Arif, Natalie Bowen, Jennifer Bole, Gary Kunzi, Daniel Bruce, Ahmed Abdelgawad, and Kumar Yelamarthi. "Smart car: An IoT based accident detection system." In 2018 IEEE Global Conference on Internet of Things (GCIoT), pp. 1-5. IEEE, 2018.

[6]   Thentral, T. M., S. Raghav Manikandan, T. R. B. Ramanathan, S. Prasanth, A. Geetha, and S. Usha. "Smart Car Interior Cleaning Robot (STRIC)." International Journal of Electrical Engineering & Technology 11, no. 3 (2020).

[7]   Mudaliar, Shruthi, Shreya Agali, Sujay Mudhol, and C. Jambotkar. "IoT based smart car parking system." Int J Sci Adv Res Technol 5, no. 1 (2019): 270-272.

[8]   Alsafery, Wael, Badraddin Alturki, Stephan Reiff-Marganiec, and Kamal Jambi. "Smart car parking system solution for the internet of things in smart cities." In 2018 1st International Conference on Computer Applications & Information Security (ICCAIS), pp. 1-5. IEEE, 2018.

[9]   Huang, Shih-Chia, Bo-Hao Chen, Sheng-Kai Chou, Jenq-Neng Hwang, and Kuan-Hui Lee. "Smart car [application notes]." IEEE Computational Intelligence Magazine 11, no. 4 (2016): 46-58.

[10]   Xu, Qinyi, Beibei Wang, Feng Zhang, Deepika Sai Regani, Fengyu Wang, and KJ Ray Liu. "Wireless ai in smart car: How smart a car can be?." IEEE Access 8 (2020): 55091-55112.

[11]   Varaiya, Pravin. "Smart cars on smart roads: problems of control." IEEE Transactions on automatic control 38, no. 2 (1993): 195-207.

[12]   Arena, Fabio, Giovanni Pau, and Alessandro Severino. "An overview on the current status and future perspectives of smart cars." Infrastructures 5, no. 7 (2020): 53.

[13]   Corona, Daniele, and Bart De Schutter. "Adaptive cruise control for a SMART car: A comparison benchmark for MPC-PWA control methods." IEEE Transactions on Control Systems Technology 16, no. 2 (2008): 365-372.

[14]   Wu, Zhaohui, Yanfei Liu, and Gang Pan. "A smart car control model for brake comfort based on car following." IEEE transactions on intelligent transportation systems 10, no. 1 (2008): 42-46.

[15]   Hasan, Md Nazmul, S. M. Didar-Al-Alam, and Sikder Rezwanul Huq. "Intelligent car control for a smart car." International Journal of Computer Applications 14, no. 3 (2011): 15-19.

[16]   Sun, Jie, Yongping Zhang, and Kejia He. "Providing context-awareness in the smart car environment." In 2010 10th IEEE International Conference on Computer and Information Technology, pp. 13-19. IEEE, 2010.

[17]   Sundaram, B. Ramya, Shriram K. Vasudevan, E. Aravind, G. Karthick, and S. Harithaa. "Smart car design using RFID." Indian Journal of Science and Technology 8, no. 11 (2015): 1-5.

[18]   Alsafery, Wael, Badraddin Alturki, Stephan Reiff-Marganiec, and Kamal Jambi. "Smart car parking system solution for the internet of things in smart cities."

In 2018 1st International Conference on Computer Applications & Information Security (ICCAIS), pp. 1-5. IEEE, 2018.

[19] Xu, Jin, Guang Chen, and Ming Xie. "Vision-guided automatic parking for smart car." In Proceedings of the IEEE Intelligent Vehicles Symposium 2000 (Cat. No. 00TH8511), pp. 725-730. IEEE, 2000.

[20] Priceman, Saul J., Stephen J. Forman, and Christine E. Brown. "Smart CARs engineered for cancer immunotherapy." Current opinion in oncology 27, no. 6 (2015): 466.

[21] Collier, W. Clay, and Richard J. Weiland. "Smart cars, smart highways." IEEE spectrum 31, no. 4 (1994): 27-33.

[22] Gupta, Aniket, Sujata Kulkarni, Vaibhavi Jathar, Ved Sharma, and Naman Jain. "Smart car parking management system using IoT." Am. J. Sci. Eng. Technol 2, no. 4 (2017): 112-119.

[23] Wang, Fei-Yue, Daniel Zeng, and Liuqing Yang. "Smart cars on smart roads: an IEEE intelligent transportation systems society update." IEEE Pervasive Computing 5, no. 4 (2006): 68-69.

[24] Wang, Fei-Yue, Daniel Zeng, and Liuqing Yang. "Smart cars on smart roads: an IEEE intelligent transportation systems society update." IEEE Pervasive Computing 5, no. 4 (2006): 68-69.

[25] Kotb, Amir O., Yao-Chun Shen, Xu Zhu, and Yi Huang. "iParker—A new smart car-parking system based on dynamic resource allocation and pricing." IEEE transactions on intelligent transportation systems 17, no. 9 (2016): 2637-2647.

[26] Atif, Yacine, Jianguo Ding, and Manfred A. Jeusfeld. "Internet of things approach to cloud-based smart car parking." Procedia Computer Science 98 (2016): 193-198.

[27] Jurgen, Ronald K. "Smart cars and highways go global." IEEE spectrum 28, no. 5 (1991): 26-36.

[28] Healey, Jennifer, and Rosalind Picard. "SmartCar: detecting driver stress." In Proceedings 15th International Conference on Pattern Recognition. ICPR-2000, vol. 4, pp. 218-221. IEEE, 2000.

[29] Atiqur, Rahman, and Yun Li. "Automated smart car parking system using raspberry Pi 4 and iOS application." International Journal of Reconfigurable and Embeded Systems (IJRES) 9, no. 3 (2020): 229-234.

[30] Soultana, Abdelfettah, Faouzia Benabbou, and Nawal Sael. "Context-awareness in the smart car: study and analysis." In Proceedings of the 4th International Conference on Smart City Applications, pp. 1-8. 2019.

YOUR KNOWLEDGE HAS VALUE

- We will publish your bachelor's and
  master's thesis, essays and papers

- Your own eBook and book -
  sold worldwide in all relevant shops

- Earn money with each sale

Upload your text at www.GRIN.com
and publish for free